Mathenglish

Vorwort

Liebe Schülerinnen und Schüler, liebe Eltern und Erziehungsberechtige, liebe Lehrkräfte,

das vorliegende Übungsbuch stellt ein Novum in der Schulbuchliteratur dar. Es soll zum einen dem Einüben von rechnerischen Grundfertigkeiten helfen, zum anderen ist jedes Kapitel ein Übungseinheit für Englisch. Es ist weder ein reines Übungsbuch für den Englischunterricht, noch ein reines Übungsbuch für den Matheunterricht. Vielmehr dient es dem effizienten Üben von Grundrechenarten und bietet ein „Sprachbad", in dem die Textaufgaben und die Aufgabenstellungen selbst sowie auch grammatikalische Erklärungen in Englisch gehalten sind. Schüler können damit ihr Englisch verbessern, in dem sie mit Aufgabenstellungen in Englisch konfrontiert sind, gleichzeitig trainieren sie die Grundrechenarten und fördern damit ihre mathematischen Kompetenzen.

Das Buch eignet sich für :

- das Wiederholen von Mathe und Englisch zu Hause (z.B. als unterstützendes und wiederholendes Lernen),

- für Vertretungsunterricht,

- für bilingualen Mathe-Unterricht,

- für Lernschleifen zu Hause in den Ferien oder vor Klassenarbeiten,

- für gleichzeitiges Sprach- und Rechentraining!

Das Buch ist auf die Lehrplaninhalte der Klassenstufe 5 bis 7 abgestimmt und eignet sich besonders für Schüler dieser Klassenstufen!

Viel Spaß und einen hohen Lernertrag in Mathe und Englisch wünschen Ihnen,

Clemens Kaesler & Ruben Kaesler

(Frankenthal, März 2020)

www.powerlerner.de

1. Addition / Word-field: food / Simple Present

1.1 Addition / Meat / Simple Present

Vocabulary:

Meat:	Please translate the words into German:
bacon	
beef	
chicken	
ham	
lamb	
pork	
sausage roll	
turkey	
sausages	

Grammar: Simple Present

Verwendung: Das Simple Present ist eine Zeitform, mit der Handlungen in der Gegenwart ausgedrückt werden, die regelmäßig oder wiederholt stattfinden.

Bildung des Simple Present:
Es gilt die Regel:

He, she, it...das –s muss mit.

(Die Verben haben ihre Grundform, bei der 3. Person (he, she, it) wird an das Verb ein –s angefügt.

Examples:
- Tom eats a sausage every day.
- Lisa likes beef.
- I like turkey at Christmas.

Practise time:

1.) John isst 2 Würste, Lisa isst 4 Würste und Tom isst sogar 5 Würste.
Translate into English!
How many sausages do the children it in all?

2.) There are fifty sausage rolls in a box. Tom brings another box with 67 sausage rolls.
Underline all verbs in the text! (verb = „Tunwort")
How many sausage rolls are there in all?

3.) Tom eats five slices of bacon on Monday, six on Tuesday and seven on Thursday. How many slices of bacon does he eat in all?
Translate into German and caculate the result!

4.) Jerry´s family loves sausage rolls for breakfast. On Monday they eat 6 rolls, on Tuesday they 7 rolls, on Wednesday they eat 5 roll and on Thursday they eat another 9 rolls. On Friday they usually don´t eat meat!

How many rolls have they eatem at the end of the week?

1.2 Addition / Dairy Products / Simple Present - Verneinung

Vocabulary:

Dairy Products	Please translate the words into German:
butter	
cream	
cheese	
milk	
full-fat milk	
semi-skimmed milk	
yoghurt	

Grammar: Negation (Verneinung)

Example 1:
Tom eats five yoghurts per week! (wrong! → 7 yoghurts)
   → Tom **doesn´t eat** five yoghurts, he eats seven yoghurts).

Example 2:
I drink a liter of full-fat milk! (wrong! → semi-skimmed milk)
I **don´t drink** a liter of full-fat milk, I drink a liter of semi-skimmed milk.

Practise time:

I. English
1.) Verneine folgende Sätze:

a) Tom eats a pound of butter!

b) You like full-fat milk!

c) Tim drinks semi-skimmed milk!

II . Maths - Solve the following word-problems

1.) A cow gives milk every day. On Monday she gives 12 litres. On Tuesday she gives 18 litres, on Wednesday she gives 21 litres.

Is it correct: <u>She gives 52 litres in all?</u>

Verneine den Satz und schreibe die richtige Lösung:

2.) Jenny loves whipped cream (dt. Schlagsahne). For a birthday party she buys six cans of cream. Her mom doesn´t know that and she also buys 8 cans of cream for the birthday cake.

Is it correct: <u>There are 16 cans of cream for the party?</u>

Verneine den Satz und schreibe die richtige Lösung:

3.) Robert loves apples. He buys one pack of apples with 3 apples and one with 7 apples.

Is it correct: <u>Robert buys 20 apples?</u>

Verneine den Satz und schreibe die richtige Lösung:

4.) Linda bakes a cake with flour, sugar and butter. She uses 30g of butter, 20g of sugar and 70g of flour.

Is it correct: <u>The cake weighs 120 g?</u>

Verneine den Satz und schreibe die richtige Lösung:

1.3 Addition / word-field: fruits / Simple Present – Asking questions?

Vocabulary:

fruits	Please translate the words into German:
apple	
orange	
banana	
pineapple	
pear	
blackberry	
strawberry	
raspberry	
cherry	

Grammar: ***Questions in English:***

Für Fragen im Simple Present in Englisch benötigt man meist ein Hiflsverb!

Example:
Tom loves cherry-cake!
→ **Does** Tom love cherry-cake?
Jenny eats a banana!
→ **Does** Jenny eat a banana?

Das Fragewort wird voran gestellt:

- **What does Jenny eat? (Was isst Jenny?)**
- **Where does Tom eat? (Wo isst Tom?)**

Practise time:

I. English:

1.) Translate into German:

English	German
Where does Jenny eat the banana?	
What does Tom like?	
When does Timmy eat the strawberry cake?	
Who eats all the apples?	

II. Maths

2.) Solve the following word-problems

a.) There are 56 apples in one box! In another box, there are 68 apples!

How many apples are in both boxes?   (dt.: both – beide)

b.) For a fruit salad you need 2 bananas, 3 apples, 12 cherries, 23 raspberries!
How many pieces of fruit do you need for the fruit salad?
(dt.: need – brauchen)

c) For a soup, you need 2 carrots, 3 sticks of celery, one leek and 4 potatoes. How many vegetables do you need in total?

d) Tom eats 2 chocolate bars. One weighs 20 g, the other one weighs 30 g. How much chocolate bar does Tom eat in grams?

e) Tom picks a basket full of strawberries. There are 28 strawberries in the basket. His friend tim also picks a basket full of strawberries. There are 23 strawberries in his basekt. How many strawberries are in both baskets?

1.4 Addition / Practise Time

1.) Do the following exercises.

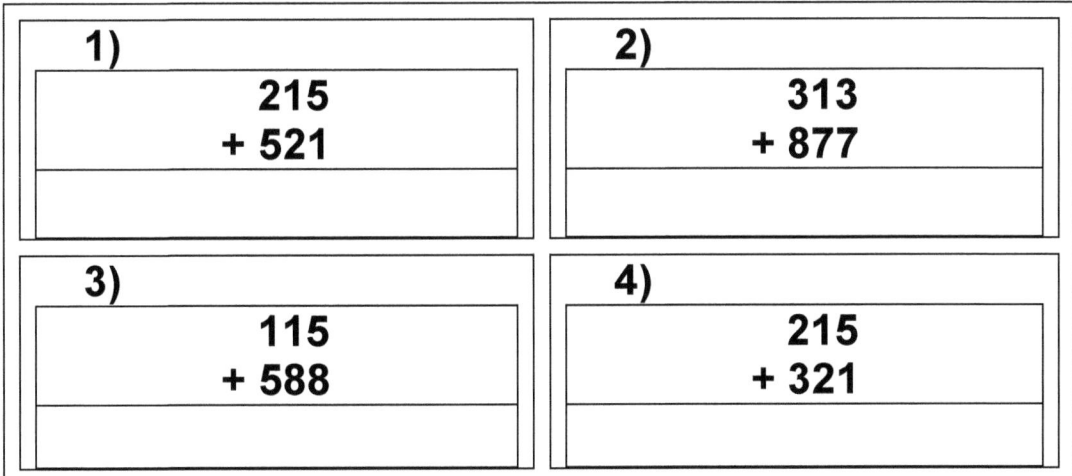

1)	2)
215 + 521	313 + 877

3)	4)
115 + 588	215 + 321

I. English:

Now write your solutions in figures (schreibe die Zahlen aus)

Example:

253 → two hundred and fifty three

a) 154	
b) 89	
c) 474	
d) 111	

II. Maths

1.) Find the missing numbers!

2182
+
4856

1429
+
3961

2.) Round to the nearest:
(Deutsch: to round - runden)

Example:
   a) 2686

*10:___2690_____   100:___2700_____   1000:__3000_*

   b) 7885

10_____   100_____   1000_____

   c) 2225

10_____   100_____   1000_____

   d) 9485

10_____   100_____   1000_____

3.) Find the missing number:

1522
+
3966

7422
+
9966

Now round the results and write in English for the results:

Example:
3966 is about 4000!

4.) Find the missing number:

1922
+
5866

1422
+
93911

Now round the results and write in English for the results:

2. Subtraction

2.1 Subtraction / Word-field: food / Simple Past

Vocabulary:

fast food	Please translate the words into German:
hamburger	
cheese	
french fries	
pizza	
slice of pizza	
pasta	
fish & chips	
breakfast	
lunch	
dinner	
scrambled eggs	

Grammar: Simple Past

Die Zeitform „Simple Past" drückt Handlungen in der Vergangenheit aus, die einmalig oder wiederholt stattfinden und komplett abgeschlossen sind.

Bildung:

Das Simple Past wird gebildet in dem an das Verb die Endung –ed angefügt wird.

Example:

- **Tom plays football.**
- **Tom play_ed_ football.**

Wichtig:

Es gibt viele unregelmäßige Verben, die im Simple Past anders gebildet werden. Diese musst du alle auswendig lernen.

Here are some important ones

Simple Present	Simple Past
to eat	ate
to buy	bought
to have	had
to cut	cut
to be	was / were

Examples:
- He is ill. → He was ill.
- They are very happy. → They were very happy.
- He buys a book. → He bought a book.

Practise time:

I. English

1.) Put into the simple past:

Simple Present	Simple Past
Tom likes pizza!	Tom lik**ed** pizza!
Jenny cooks scrambled eggs!	
Peter has pasta for lunch!	
Timmy buys a hamburger with french fries.	

II. Maths - Solve the following word-problems (Subtraction)

1.) There were 16 hamburgers for the birthday party! The kids ate 9 of them. How many were left?

2.) Tom ordered 5 pizzas for the film night with his friends. Each pizza was cut into 12 slices. They ate 48 slices! How many were left?

3.) McDonalds in Frankenthal sells 1689 burgers (hamburgers and cheeseburgers) a day. 856 of them are cheeseburgers. How many hamburgers does McDonalds sell per day?

a) Put the text into the simple past! (to sell - sold [Deutsch: verkaufen]).

b) What is the result of your calculation?

4.) McDonalds in Hamburg sells 2981 burgers (hamburgers and cheeseburgers) a day. 1856 of them are cheeseburgers. How many hamburgers does McDonalds sell per day?

What is the result of your calculation?

2.2 Subtraction / Word-field: beverages / Comparison of Adjectives

Vocabulary:

beverages	Please translate the words into German:
apple juice	
orange juice	
lemonade	
coke	
milk	
milk-shake	
cup of coffee	
cup of tea	
water	
sparkling water	

Grammar: Vergleich von Adjektiven
             genauso... wie = as... as

Example: Terry is as strong as Bill.
             nicht so... wie = not as... as / not so ... as
             John is not as strong as Jim.

Practise time:

I. English

1.) Translate into German

English	German
Tom is as tall as Lisa!	
Dagobert Duck is as rich as Claas.	
Motorbikes are as fast as sports cars.	

English	German
	Tomaten sind so rot wie Kirschen.
	Der Turm ist so groß wie die Kirche!
	Das Haus ist so klein wie ein Wohnwagen (engl. caravan).

II. Maths

1.) Solve the following word-problems

a) Tina ate as many chewing gums as tom. Together they ate 12 chewing gums. There were 36 chewing gums in one package at the beginning. How many chewing gums are left? How many did Tina eat?

b) A lego tower consists of 157 block. 112 block are red, the rest is green. How many green blocks are in the tower?

c) A farmer has 24 pumpkins in his garden. For halloween, he carves 10 of them. How many has he left to eat?

d) The farmer also has some chicken. In the morning, he collects 20 eggs. He cooks breakfast with 4 of them. How many has he left for the next day?

e) The farmer has 20 chickens, but one night, a fox steals 8 of them. How many has he left?

3. Multiplication

3.1 Multiplication / Word-field: Sports / Adjectives

Vocabulary:

Sports	Please translate the words into German:
football / soccer (AE)	
running shoes	
hockey stick	
racquet	
tennis racquet	
table tennis	
swimming	
riding	
cycling	
hiking	
athletics	

Grammar: Steigerung von Adjektiven

Adjektive werden so gesteigert:
big – bigger – the biggest
fast – faster- the fastest
small – smaller – the smallest

Bei mehr als zwei Silben wird so gesteigert:
important – more important – most important
difficult – more difficult – most difficult

Practise time:

I. English

1.) Translate into English

English	German
	Tom ist schneller als Lisa. (big)
	Tom ist der größte in Leichtathletik.
	Claus ist der schnellste.

English	German
Maths is more difficult than English	
The movie is more exciting than the book.	
Tim´s shirt is dirtier than Tom´s shirt.	
The book is funnier than the movie.	

II. Maths

1.) Tom loves running. He runs 6 laps (dt. Runden) in the stadium. One lap is 400 metres.

a) How many meters does he run in all?
b) How many metres does he run if he runs 9 laps?

2.) Tim likes hiking. He walks 12 kilometers each Sunday. How many kilometers does he walk on 8 Sundays?

3.) Sharon buys three pairs of running shoes. Each pair costs 125 Euros.
a) How much money does Sharon spend on running shoes?

b) Peter also buys new running shoes. His pair of running shoes costs 159 Euros. Which shoes are more expensive? Peter´s or Sharon´s?

3.2 Multiplication / Mental Maths / Multiply 1 digit by 1 digit

(Mental Maths → Deutsch: Kopfrechnen)
(Multiply 1 digit by 1 digit - Einstellige Zahlen multiplizieren)

Vocabulary:

If you **multiply** two digits, the result is called "product".
(Wenn du zwei Zahlen multiplizierst, heißt das Ergebnis „Produkt".)

If you **add** two digits, the result is called "sum".
(Wenn du zwei Zahlen addierst, heißt das Ergebnis „Summe".)

Practise time:

Find the product!

$3 \times 5 =$	$3 \times 7 =$
$7 \times 9 =$	$7 \times 8 =$
$3 \times 8 =$	$4 \times 6 =$
$5 \times 2 =$	$4 \times 3 =$
$7 \times 8 =$	$5 * 9 =$
$9 \times 2 =$	$7 \times 6 =$

3.3 Multiplication / Mixed Word-Problems

The following word problems can be solved by mental multiplication. But it is very helpful, if you write out the answers like in the example.

(Deutsch: mental multiplication – Multiplizieren im Kopf)

Example:

There are 9 apples in each box. How many apples are in 6 boxes?

9 apples x 6 boxes = 54 apples

54 apples are in 6 boxes

Practise time:

1.) There are 9 apples in each box. How many apples are in 6 boxes?

2.) Each child has 4 tickets. If there are 6 children, how many tickets are there in total?

3.) There are 9 erasers in each box. How many erasers are in 9 boxes?

4.) There are 6 cards in each box. How many cards are in 9 boxes?

5.) Juan has 8 boxes of eggs. Each box holds 6 eggs. How many eggs does Juan have?

3.4 Distributive property of mulitplication (Zerlegung)

Multiplying bigger numbers is not easy. You can help yourself, if you split one factor in two parts.

(to split something – Deutsch: etwas aufteilen → 24 is split in 20 and 4)

Example:

4 * 24 = (4 * 20) + (4 * 4)   // 24 is split up in 20 and 4

Practise Time:

1.) Rewrite the calculations like in the example:

7 * 14 = (7 * 10) + (7 * 4) = 98
3 * 13 =
8 * 19 =
5 * 15 =
3 * 12 =
6 * 16 =
7 * 13 =
9 * 19 =
4 * 17 =
3 * 12 =
5 * 14 =
9 * 13

4. Division

4.1 Division / Word-field: Garden and Nature / Simple Past

Vocabulary:

Garden and Nature	Please translate the words into German:
tree	
lawn	
bush	
pond	
lake	
to hike	
hiking trail	
forest	
jungle	
to plant	
spring	
harvest	

Practise Time:

I. English

Transfer the tense of following sentences from the Simple Present into the Simple Past.

Simple Present	Simple Past
John plants an apple tree.	
Joana does a hiking tour in the black forest every Sunday.	
Tina likes working in the garden.	
Tom walks around the lake in summer.	
Mr Smith plants salad in his garden in spring.	
Mrs Smith harvests the tomatoes in summer.	

II. Maths

1.) Mr Smith harvests 36 tomatoes in his garden. He has got 9 tomatoe plants.
How many tomatoes can Mr Smith harvest from each plant on average?

Vocabulary:
on average – im Durchschnitt

2.) John helped planting trees. By the end of the week, he planted 165 trees. He worked from Monday to Friday.
How many trees did he plant on a day?

3.) Joana did a four day hiking tour in the lake district. The trip was 48 kilometers.
How many kilometers did she walk per day?

4.) Mr Smith harvested 81 cucumbers in his garden last summer. He had 9 cucumber plants. How many cucumbers did he harvest from each plant?

5. Division without remainders - Work out the answer to these calculations. (Schrifliche Division ohne Rest).

a.) 445 : 5 = _____

b.) 504 : 6 = _____

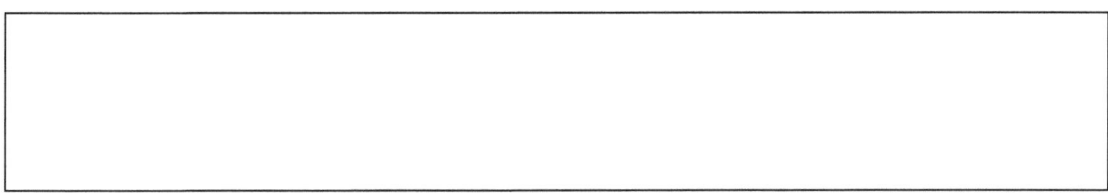

c.) 360 : 8 = _____

d.) 2472 : 6 = _____

5. Calculating with money

(One Dollar Note (1 Dollar Note USA))

(British Pound   20 £, Währung von Großbritannien)

Grammar:

Difference „How much…." // „How many…."..

`Much' should be used with uncountable nouns, while 'many' should be used with countable nouns.

Pracitise time:

1.) Addition of big numbers:

102.500,- €	**102.500,- €**
+108.000,-€	**-96.000,-€**
102.500,- €	**500.000,- €**
+108.000,-€	**-178.000,-€**
254.500,- €	**78.500,- €**
+110.000,-€	**-10.500,-€**
2.500,- €	**170.500,- €**
+1.700,-€	**-108.500,-€**
730.800,- €	**102.500,- €**
+125.200,-€	**-100.000,-€**

2.) Addition and subtraction of three numbers:

120.500,- € + 117.200,- € + 6.200,- €

132.500,- € - 108.000,- € - 5.200,- €

22.700,- € + 50.000,- € + 5.200,- €

187.700,- € - 118.000,- € - 4.300, -€

102.500,- € +108.000,- € + 7.200,- €

125.500,- € - 118.000,- € - 5.600,- €

4.1472.500,- € + 3100.000,- € + 65.100,- €

145.500,- € -123.200,-€ -22.300,-€

3.) Solve the following word problems

a.) Tim gets 10,- $ pocket money per week. Each school-day he buys a bottle of milk at school (0,80 $). On Mondays he also buys his favourite magazine on computer games (2,80 $).
How much money is left at the end of the week?

New vocabulary:
- *pocket money = Taschengeld*
- *favouritemagazine = Lieblingszeitschrift*
- *end oftheweek = am Ende der Woche*

b.) Mark has 700$ on his bank account. In January he spends 100$ on a new bike, 297$ on a new smartphone and 159 $ on software.

How much money does he have left at the end of the month?

New vocabulary:
- *bank account= Bankkonto*

4.) Kevin earns 5.50 $ for delivering a Pizza in his town and 7.50 $ for delivering Pizza in a town nearby. He delivered 3 Pizzas in her town today, and 2 Pizzas in the town nearby.

New vocabulary:
- deliver= ausliefern
- earn= verdienen
- How much money did he earn?

6. Scale units

6.1 Measuring length (metric system)

Typical scale units for measuring thelenghtof an object are:

km = kilometers (Kilometer)
m = meters (Meter)
centimeters (Centimeter)

1 meter is 100 cm
1 cm is 100 mm

Rechne um in die fehlenden Maßeinheiten (Transfer into the missing scale unit)

km	m	cm
1	1000	100.000

km	m	cm
2.5	2.500	250.000
12		
18		

km	m	cm
0,1	100	10.000
		500.000
		1.300.000

6.2 Measuring Weight

1.) Complete the following table:

t (ton)	kg (kilogramm)	g (gramm)
5		
0.1	100	
		1.300.000
7.5		
		1000
	200	
3		
	2000	2000000
1.2		
	500	
70		50000
	10	
	25.5	

2.) Tom goes shopping every Monday. He buys a pack of potatoes (1.5 kg), 10 carrots (0.8 kg), two packs of flour (1 kg per pack), six bars of chocolate (120g each!).

New vocabulary:
- *flour = Mehl*
- *bar of chocolate = Tafel Schokolade.*

How much weighs Tom´s shopping bag?

3.) Lisa goes to the grocery store. She buys two liters of milk, 5 kg of potatoes, 200 grams of flour and a turkey which weighs 3 kgs.

New vocabulary:
- Turkey = Truthahn

How much does her shopping bag weigh?

4.) A truck transports kitchen furniture. It transports tables which weigh 2 tons together,a refrigerator which weighs 500 Kg and an oven which weighs 300 Kg. The truck itself weighs 20 tons.

7. Uhrzeit lernen / Learn to read the clock

7.1 Full hours

Im Englischen geht die Uhrzeit immer bis 12 Uhr. Ist es Nachmittags bis Mitternacht ergänzt

LONDON   NEW YORK   TOKYO   MOSCOW

man die Uhrzeit mit "pm" [gesprochen pi ähm], ab Mitternacht bis 12. Uhr ergänzt man mit "am" [gesprochen ei ähm]. Die Ergänzung macht man nur um Verwechslungen vorzubeugen, es ist z.B. klar, dass die Schule morgens um 09.00 Uhr (9 o´clock) beginnt, dann muss man nicht noch "am" ergänzen.

Vollstunde werden immer mit o´clock (of the clock) ergänzt:

Example:

German	English
12.00 Uhr	12 o´clock
16.00 Uhr	4 o´clock am (not 16 o´clock!)

1.)

Do the following exercise (mache die folgende Übung):
Transfer the digital time, into the English way of saying the time.

Digital	Write in English	Show the time in the watch:
01.00 Uhr		

17.00 Uhr		
05.00 Uhr		
21.00 Uhr		
23.00 Uhr		

7.2 Quarter of an hour

"Viertel nach" wird mit „quarter past" ausgedrückt. „Viertel vor" wird mit „quarter to" zur nächsten vollen Stunde ausgedrückt.

Examples:

German	English
02.15 Uhr	quarter past two
16.15 Uhr	quarter past four
16.45 Uhr	quarter to five
05.45 Uhr	quarter to six

Digital	Write in English	Show the time in the watch:
01.15 Uhr		
17.45 Uhr		
05.15Uhr		
21.45 Uhr		
22.45 Uhr		
06.45 Uhr		

2.) Tom gets up at 6.45 in the morning. He puts on his clothes (3 minutes) goes to the bathroom (8 minutes) and has breakfast with his brother Joe and his dad (16 minutes). After that he leaves the house!

a) What time is it, when he leaves the house?

b) Tom has to walk to the bus stop. That takes him 6 minutes. The school bus goes at 07.38 at his bus stop. Before he walks to the bus stop he plays with his dog Charly. How many minutes has he got to play with his dog before he has to walk to the bus stop?

3.)

	Show the time in the watch
Quater past five!	
Half past four!	
Five to nine!	

Five past six	
Half past eight	
Quater to nine	
Twenty past nine	
Quater to twelve	
Quater to six	
Half past seven	

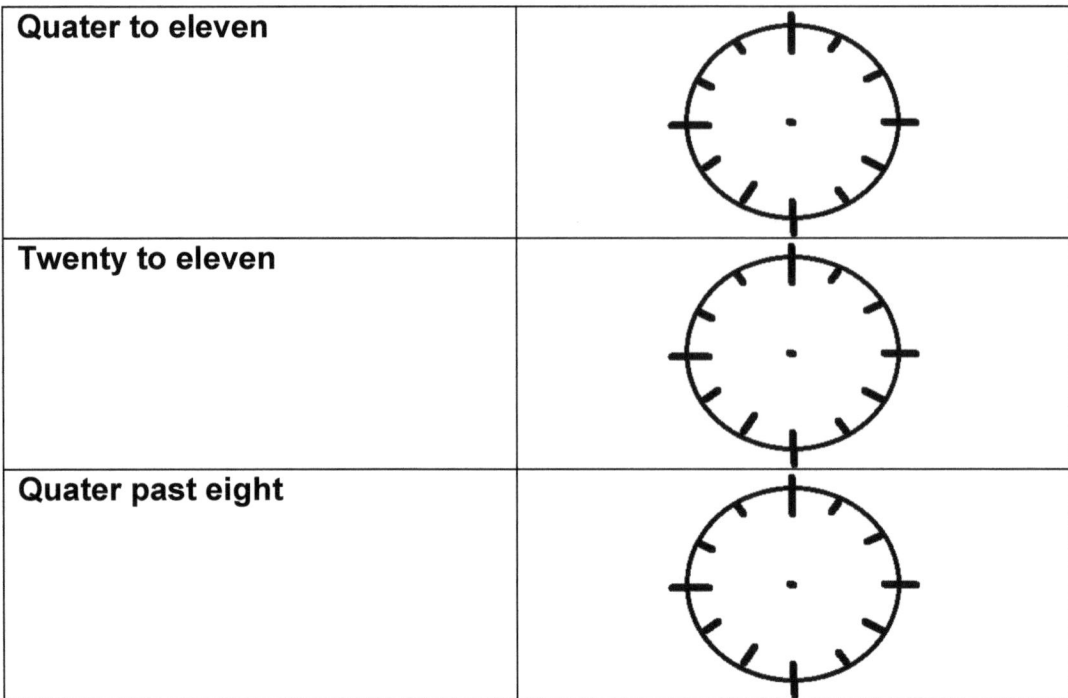

Quater to eleven	
Twenty to eleven	
Quater past eight	

Liebe Eltern und Erziehungsberechtigte, liebe Lehrkräfte,

wir hoffen, dass das Buch Ihre Kinder beim Lernen von Mathe und Englisch unterstützt hat und freuen uns über ein Feedback:

clemens.kaesler@gmail.com.

Clemens Kaesler
Ruben Kaesler

www.powerlerner.de

© 2020
Herstellung und Verlag: BoD – Books on Demand, Norderstedt
ISBN: 978-3-7519-0242-7